PERENNIAL MATH

Tournament Tests 2

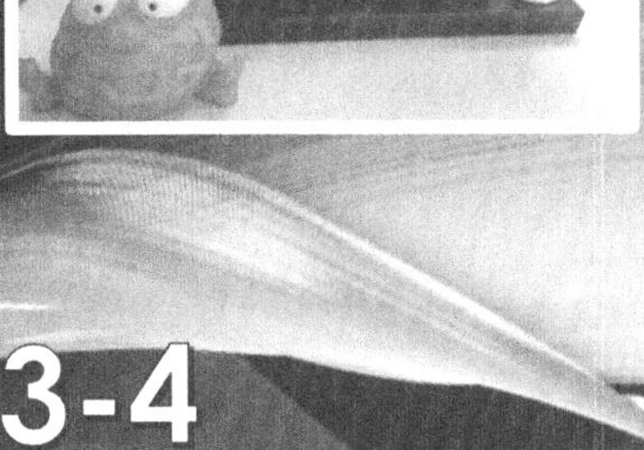

Grades 3-4

PERENNIAL MATH TOURNAMENT TESTS

These are a collection of tests from past Perennial Math Tournaments. Each tournament has two rounds of testing. Each student takes an Individual Test during the first round and then students combine with their teammates to take a Team Test during the second round. The solutions included in this book are only a short sample of the many possible solutions that could be used to solve each problem.

INDIVIDUAL TESTS GRADES 3-4

30 minute practice tests for individual students

Grade 3-4 Individual Event (30 Minutes)

1) If Andrea puts her jellybeans in piles of 4 she has 2 jellybeans left over. If she puts her jellybeans in piles of 5 she has 3 jellybeans left over. What is the least amount of jellybeans Andrea can have?

2) This puzzle contains the numbers 5-13. Each column and row add up to the number outside the boxes. Put the correct numbers in ach box to complete the puzzle without repeating any numbers. What number should go in the middle box?

			24
			28
			29
27	28	26	

3) What is the perimeter of the shaded region?

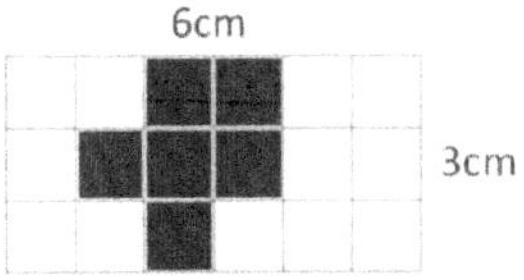

4) Jennifer bought 10 gumballs at 25 cents each. If she sells all the gumballs for the same price, how much will she have to sell each gumball for in order to make a $1 profit?

5) Find the average of the numbers between 1 and 100 that end in 6.

6) A dog buries 10 bones on Sunday, 20 on Monday, 30 bones on Tuesday, and 40 bones on Wednesday. If he continues to bury 10 more bones a day, how many bones will he have buried at the end of a week?

7) The original price of a new tennis racquet is $140.00. If the racquet is marked down 10%, what is the new price of the racquet?

8) There are twice as many boys as girls in the gym. If there are 207 total children, how many girls are there?

9) Kim made 48 pumpkin cookies. She put black hats on half of the cookies, teeth on one third of the cookies, and the remaining cookies were blank. How many cookies were blank?

10) There are 42 students on the football team. Fourteen of the players are out wit the flu. What fraction of the team is left to play?

11) $12 + 23 + 34 + 45 + 55 + 66 + 77 + 88 =$

12) What is the sum of the first 8 prime numbers?

13) $256 \div 4^2 =$

14) $17 \times 27 =$

15) If the area of a square is $25cm^2$, what is the perimeter of the square?

INDIVIDUAL ANSWER DOCUMENT (Grades 3-4)

SCORE

Team # __________ School ______________________________

Student Name______________________________________

Student Number________________

1. ________	6. ________	11. ________
2. ________	7. ________	12. ________
3. ________	8. ________	13. ________
4. ________	9. ________	14. ________
5. ________	10. ________	15. ________

PERENNIAL MATH

Grade 3-4 Individual Event Solutions

1) Make a chart with multiples of 4 adding 2 and compare with multiples of 5 adding 3. When the number is the same for both, that is the answer. The least number possible is 18 jellybeans.

Piles of 4 with 2 jellybeans left over	6	10	14	18
Piles of 5 with 3 jellybeans left over	8	13	18	

2) Complete the top row by adding 8 to get a sum of 24. Then complete the last column by adding 5 in the middle to get a sum of 26. The numbers left to choose from are {7, 9, 11, 12}. After you arrange these numbers to get the sums of the columns and rows, 11 is the middle number.

6	10	8	24
12	11	5	28
9	7	13	29
27	28	26	

3) Since the dimensions of the original shape is 6cm x 3cm, then length of one side of each of the small squares is 1cm. Therefore, the perimeter of the shaded region is 12cm.

4) Jennifer originally spent 10 x .25 or $2.50. In order to make a $1 profit he must sell each lollipop for .35 to make .35 x 10 or $3.50

5) Add up the numbers that end in 6: 6 + 16 + 26 + 36 + 46 + 56 + 66 + 76 + 86 + 96 = 510. Now divide 510 by 10 and the average is 51.

6) 10 + 20 + 30 + 40 + 50 + 60 + 70 = 280 bones

7) The original price of a new racquet is $140.00. $140 x 10 % is $14. Subtract the 10% to find the new price of the sled: $140 -$14 = $126

8) There are twice as many boys as girls in the gym. If there are 207 total children, how many girls are there? 207/3 = 69. There are 69 girls and 138 boys.

9) Kim made 48 pumpkin cookies. Half (24) of the cookies had hats, 1/3 (16) had teeth.
48 – 24 – 16 = 8 cookies remaining that were blank.

10) 14/42 players had the flu. This is 1/3 of the team. So 2/3 of the players were left to play.

11) Instead of doing regular addition, use smart addition and link numbers together
12 + 88 + 23 + 77 + 34 + 66 + 45 + 55 =
= 100 + 100 + 100 + 100 = 400

12) The sum of the first 8 prime numbers is 2 + 3 + 5 + 7 + 11 + 13 + 17 + 19 = 77

13) $256 \div 4^2 = 256 \div 16 = 16$

14) 17 x 27 = 459

15) Area = 25cm^2, then each side equals 5. 5cm x 4 = 20cm.

PERENNIAL MATH

INDIVIDUAL ANSWER KEY
(Grades 3-4)

Instructions for Scoring:

- ***Mark CORRECT answers by circling "O" the question number, and INCORRECT answers with an "X" . Each correct answer is worth one point. Count the circles (correct answers) and write the number in the "SCORE" box.***

CORRECT → (1.) ______

INCORRECT → X. ______

1. **18**
2. **11**
3. **12**
4. **35¢ or .35**
5. **51**
6. **280**
7. **126**
8. **69**
9. **8**
10. **28/42 or 2/3**
11. **400**
12. **77**
13. **16**
14. **459**
15. **20**

Grade 3-4 Individual Event (30 Minutes)

1) Amy had 200 pennies. She was able to double the amount and then next she increased the amount by 150 pennies. She had to give her brother half, so she was left with ____ pennies.

2) Place the numbers 1-6 in the circles so that the right side and the left side equal 12, and the bottom equals 8? What number goes in the top circle?

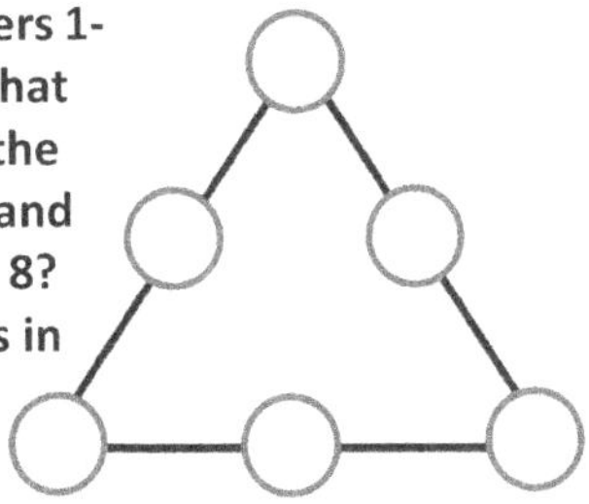

3) If this pattern were continued, how many circles would the 15th item in the series have?

4) Riddhi bought a sandwich for $4.75, a bag of chips for $1.70, and a drink for $1.75. The tax was $0.65. She gave the cashier $10.00. How much change did Riddhi receive?

5) Barry was collecting coins. He got 20 coins from his grandfather and 8 coins from his brother. His mom gave Barry 14 coins. Barry gave a third of the coins to his friend Greg. How many coins did Barry have left?

6) During the summer holidays, your brother earns extra money cleaning shutters. He cleans 6 shutters an hour and has 45 shutters to clean. How long will it take him? (Answer in decimal form)

7) Consecutive numbers are numbers that are in order such as 1, 2, 3. Find three consecutive numbers that add up to 66. What is the greatest number?

8) Your mom bought you a 1 year swimming pool pass for $390. She will pay it off in 12 equal payments. How much money will she spend in payments on the pass in 3 months?

9) Ms. Jones bought 5 packs of red markers, 8 packs of yellow markers, and 4 packs of green markers. There were 2 markers in each package. How many markers did Ms. Jones buy in all?

10) How many 3 digit numbers can you make using the digits: 2, 5, and 7 if you can repeat the digits in each number?

11) Compute 1 – 1/2 + 1/3 – 1/4. (Answer must be simplified fraction).

12) 2109
x 7

13) $1000 \div 10^2 =$

14) What is the greatest common factor of 24 and 36?

15) $2 + 2 \times 6 - 4 \div 4 =$

INDIVIDUAL ANSWER DOCUMENT
(Grades 3-4)

SCORE

Team # __________ School ______________________________

Student Name ______________________________

Student Number ________________

1. __________ 6. __________ 11. __________

2. __________ 7. __________ 12. __________

3. __________ 8. __________ 13. __________

4. __________ 9. __________ 14. __________

5. __________ 10. __________ 15. __________

PERENNIAL MATH

Grade 3-4 Individual Event Solutions

1) 200 pennies x 2 = 400 pennies. 400 + 150 = 550 pennies. 550/2 = 275 pennies.

2) Use trial and error. The top number is 6.

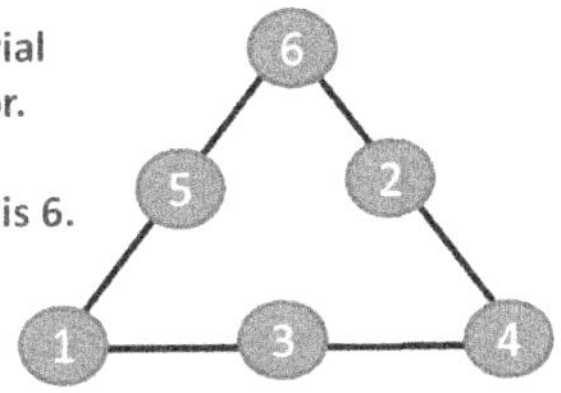

3) Each term in the sequence is 3 times the number of the term in the sequence. Therefore the 15th number in the sequence would have 15 x 3 = 45 circles.

4) Riddhi bought a sandwich for $4.75, a bag of chips for $1.70, and a drink for $1.75. The tax was $0.65. She gave the cashier $10.00. $4.75 + $1.70 + $1.75 + .65 = $8.85. $10.00 - $8.85 = $1. 15.

5) Barry got 20 coins from his grandfather + 8 coins from his brother+ 14 coins from his mom = 42 coins. Barry gave a third of the coins to his friend Greg. 42/3 = 14 coins . 42 – 14 = 28. Barry had 28 coins left.

6 He cleans 6 shutters an hour which is 1 shutter every 10 minutes. 45 shutters x 10 minutes each = 450 minutes. It will take him 7.5 hours to clean the shutters.

7) Find three consecutive numbers that add up to 66. First find the middle number by averaging the three numbers. 66/3 = 22. Since 22 is the average, the numbers are 21, 22, and 23. The greatest number is 23.

8) $390/12 = $32.50 per month. $32.50 x 3 = 97.50

9) 5 packs of red markers + 8 packs of yellow markers + 4 packs of green markers = 17 packs of markers. There were 2 markers in each package. 17 x 2 = 34 total markers.

10) How many 3 digit numbers can you make using the digits: 2, 5, and 7 if you can repeat the digits in each number? Start with 2 and make an organized list. 222, 225, 227, 252, 255, 257, 272, 275, 277. Since there are 9 numbers that can be created with 2 as the start, then there are also 9 numbers for both 5 and 7 as the beginning number. 9 x 3 = 27 possible numbers that can be made.

11) 1 – 1/2 + 1/3 – 1/4. 12/12 – 6/12 + 4/12 – 3/12 = 7/12

12) 2109 x 7 = 14,763.

13) $1000 \div 10^2 = 1000 \div 100 = 10$

14) The GCF of 36 and 48 is 12.

15) 2 + 2 x 6 – 4 ÷ 4 = 2 + (12) – (1) = 13

PERENNIAL MATH

INDIVIDUAL ANSWER KEY
(Grades 3-4)

Instructions for Scoring:

- ***Mark CORRECT answers by circling "O" the question number, and INCORRECT answers with an "X". Each correct answer is worth one point. Count the circles (correct answers) and write the number in the "SCORE" box.***

CORRECT → (1.) ________

INCORRECT → X. ________

1. 275
2. 6
3. 45
4. 376
5. 28
6. 7.5 hours
7. 23
8. 97.50
9. 34
10. 27
11. 7/12
12. 14,763
13. 10
14. 12
15. 13

Grade 3-4 Individual Event (30 Minutes)

1) Bob has 2 hamsters. Together they eat 3 grams of food twice a day. How many grams of food does Bob have to feed them in one week?

2) Using the information from the scales, how much does one square weigh?

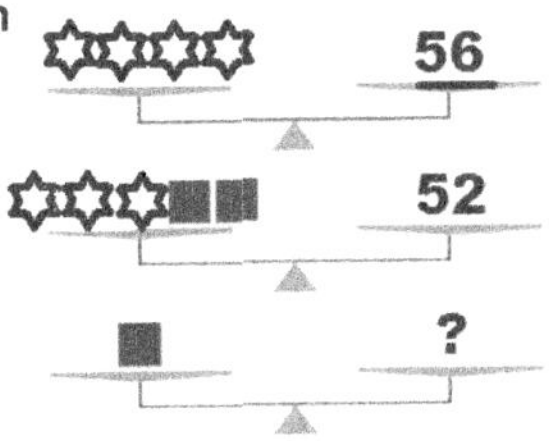

3) This 3 x 3 x 3 cube is composed of 27 unit cubes. If each face of the cube is painted as shown, how many of the 27 cubes have paint on them?

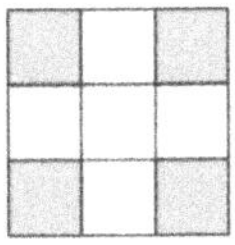

4) How many times does the digit "5" appear between 1 and 65 inclusive?

5) Misty is reading a book that has 137 pages. If she reads 8 pages a day, how many days will it take her to finish the book?

6) It is 5:30 pm. What time was it 54 hours ago? Be sure to include a.m. or p.m. with your answer.

7) Anton bought a hamburger for $8.35 and a fries for $3.25. He used a $20 bill to pay for them. How much change did Anton get?

8) Suzy and Cathy have 60 dollars together. Suzy has three times as much as Cathy. How much money does Cathy have?

9) On a 20 question math test, your teacher gives you five points for every correct answer and takes away one point for each incorrect answer. If you score 64 on the test, how many did you get correct?

10) How many rectangles can be made that have perimeter of 48 cm and sides that are whole numbers?

PERENNIAL MATH

11) What is the average of these five numbers: (24, 25, 26, 27, 28)?

12) $\begin{array}{r} 5000 \\ \underline{\times\ 90} \end{array}$

13) $9 + 2 \times 6 - 5 \times 2 =$

14) $1\frac{5}{6} \times 3\frac{3}{4} =$
(Simplify to lowest term)

15) $3328 \div 256 =$

INDIVIDUAL ANSWER DOCUMENT
(Grades 3-4)

SCORE

Team # __________ School ______________________________

Student Name ______________________________

Student Number ________________

1. ____________
2. ____________
3. ____________
4. ____________
5. ____________
6. ____________
7. ____________
8. ____________
9. ____________
10. ____________
11. ____________
12. ____________
13. ____________
14. ____________
15. ____________

PERENNIAL MATH

Grade 3-4 Individual Event Solutions

1) Three grams x 2 = 6g per day. 6g x 7 days = 42g per week.

2) Since there are 4 stars that weigh 56, then each star weighs 56/4 = 14. Substitute 4 in for each star on the next scale and 3 stars weigh 14 x 3 = 42. The two remaining squares must weigh 10 because 52-42 = 10. Each square then weighs 10/2 = 5.

3) The cube is composed of 27 unit cubes. Each corner is painted and touches the corners of the other faces. Only 8 cubes of the 27 have paint on them.

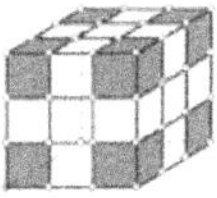

4) The digit 5 appears 17 times. Be sure to count two 5s in 55 and include the 5 in 65.
(5, 15, 25, 35, 45, 50, 51, 52, 53, 54, 55, 56, 57, 58, 59, 65)

5) 137 pages divided by 8 pages per day is 17 days with 1 page left over. So she will need to read 18 days to finish the book.

6) 54 hours = 2 days and 6 hours before 5:30 p.m. So it will be 11:30 a.m..

7) The hamburger and fries costs $8.35 + $3.25 = $11.60. $20 - $11.60 = $8.40.

8) Suzy has 3 times as much as Cathy so students can make a chart to see the relationship.

Cathy	Suzy	Total
1	3	4
2	6	8
3	9	12
4	12	16
5	15	20
6	18	24
7	21	28
8	24	32
9	27	36
10	30	40
11	33	44
12	36	48
13	39	52
14	42	56
15	45	60

9) Use a table to solve the problem.

Problems Correct	Points	Points Taken Away	Score
20	100	0	100
19	95	1	94
18	90	2	88
17	85	3	82
16	80	4	76
15	75	5	70
14	70	6	64

10) Rectangles with these dimensions can be made: (1, 23), (2, 22), (3, 21) (4, 20), (5, 19), (6, 18), (7, 17), (8, 16), (9, 15), (10, 14), (11, 13), (12, 12). There are 12 rectangles.

11) The average of the 5 odd numbers is the middle number: 26

12) 5000 x 90 = 450,000

13) [9 + (2 x 6) – (5 x 2)] = [9 + 12 – 10] = 21 – 10 = 11

14) $1\frac{5}{6}$ x $3\frac{3}{4}$ = 11/6 x 15/4 = 55/8 = $6\frac{7}{8}$

15) 3328 ÷ 256 = 13

PERENNIAL MATH

INDIVIDUAL ANSWER KEY
(Grades 3-4)

Instructions for Scoring:

- *Mark CORRECT answers by circling "O" the question number, and INCORRECT answers with an "X" . Each correct answer is worth one point. Count the circles (correct answers) and write the number in the "SCORE" box.*

CORRECT → (1.) ________

INCORRECT → X. ________

1. **42**
2. **5**
3. **8**
4. **17**
5. **18, 17.125, 17 1/8**
6. **11:30 a.m.**
7. **8.40**
8. **15**
9. **14**
10. **12**
11. **26**
12. **450,000**
13. **11**
14. **6 7/8 or 6.875**
15. **13**

Grade 3-4 Individual Event (30 Minutes)

1) How many 3 digit numbers can you make using the digits: 2, 5, and 0 if you can use the digits more than once in each number?

2) This figure is made up of cubes and some of the cubes are not visible. How many cubes did it to make this shape?

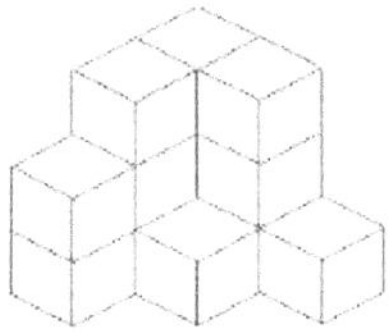

3) How many triangles are in this shape?

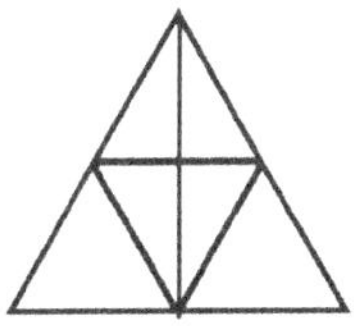

4) Jennifer and Kim have 90 books together. If Jennifer gave Kim 5 books, then they have the same number of books. How many books did Jennifer have at the beginning?

5) The sum of Brad and Carmen's ages is 36 years. Carmen's age is twice Brad's age. How old is Carmen?

6) Of all the numbers between 10 and 99 inclusive, how many times does the digit "9" appear?

7) Find the pattern and fill in the blanks.
(2, 3), (3, 5), (4, 7), (5, 9) ....
The eighth pair in this pattern would be (_____ , _____)

8) There are 19 kids including Brian standing in a line. The number of kids behind Brian is 5 times the number of kids ahead of Brian. What place is Brian in the line?

9) Jackson Elementary school has four 3rd grade classes. Each class has 25 students. At the end of the school year, 9 students transferred to another school. How many students were left in the 3rd grade?

10) Out of 15 students in Mrs. Bybee's homeroom, 6 like burgers, 9 like pizza and 3 like both. How many students in the group like neither burgers nor pizza?

PERENNIAL MATH

11) 9 - 2 x 3 + 10 =

12) 875 ÷ 25 =

13) $\frac{8}{12} \div \frac{2}{3}$ =

14) What is the average of these numbers: 15, 17, 19, 21, 23, 25, 27

15) If Scott reads 2.1 pages per minute, how many pages does he read in 1.5 hours?

INDIVIDUAL ANSWER DOCUMENT (Grades 3-4)

SCORE

Team # __________ School ______________________________

Student Name______________________________________

Student Number D - ____________________

1. ____________	6. ____________	11. ____________
2. ____________	7. ____________	12. ____________
3. ____________	8. ____________	13. ____________
4. ____________	9. ____________	14. ____________
5. ____________	10. ____________	15. ____________

PERENNIAL MATH

Grade 3-4 Individual Event Solutions

1) There are 18 three-digit numbers: 200, 202, 205, 220, 222, 225, 250, 252, 255, 500, 502, 505, 520, 522, 525, 550, 552, 555

2) This tower is made up of layers of cubes. There are 3 stacks with 3 for 9 cubes. One stack of 2 and 2 individual cubes. 9 + 2 + 1 + 1 = 13

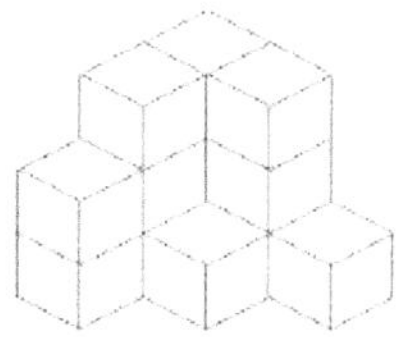

3) There are 13 triangles in the shape. 6 individual smaller triangles, and 2 large triangles.

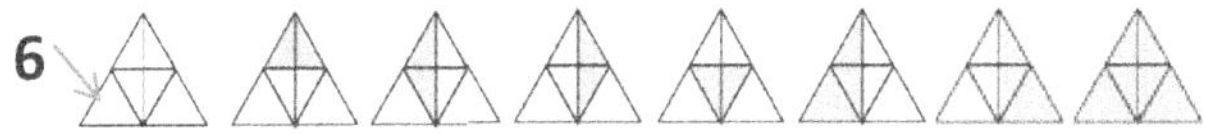

4) Jennifer has 50 books, Kim has 40 books to start with.
90 ÷ 2 = 45, Kim has 40 + 5 = 45 books, Jennifer has 50 - 5 = 45 books

5) Make a table with Brad and Carmen's age and the sum of their ages. Carmen is 24 years old, Brad is 12 years old.

Brad	Carmen	Sum 36
1	2	3
2	4	6
3	6	9
4	8	12
5	10	15
6	12	18
7	14	21
8	16	24
9	18	27
10	20	30
11	22	33
12	24	36

6) The digit 9 appears 19 times. Remember to count both nines in 99.

(19, 29, 39, 49, 59, 69, 79, 89, 90, 91, 92, 93, 94, 95, 96, 97, 98, 99)

7) (2, 3), (3, 5), (4, 7), (5, 9) .... The numbers are in a pattern such that the first term increases by 1 and the second term increases by 2. So the next pairs would be (6,11), (7,13), (8, 15) and the 8th group is (9, 17).

8) Draw this out and find where Brian is standing. Brian is 4th in line.

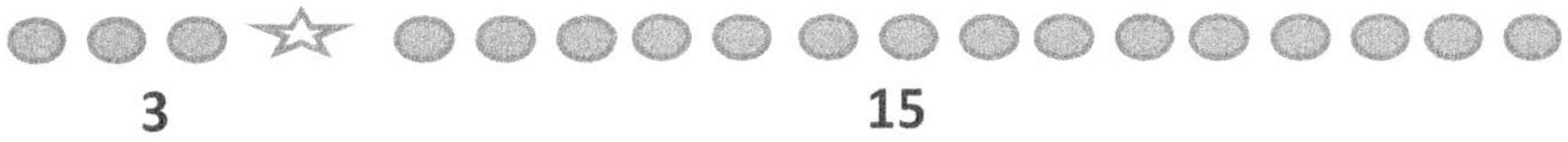

PERENNIAL MATH

Grade 3-4 Individual Event Solutions

9) 25 students x 4 classes = 100 students. 100 - 9 = 91 students left.

10) Out of 15 students in Mrs. Bybee's homeroom, 6 like burgers, 9 like pizza and 3 like both. Use a Venn Diagram to find how many students in the group like neither burgers nor pizza. 3 students like neither.

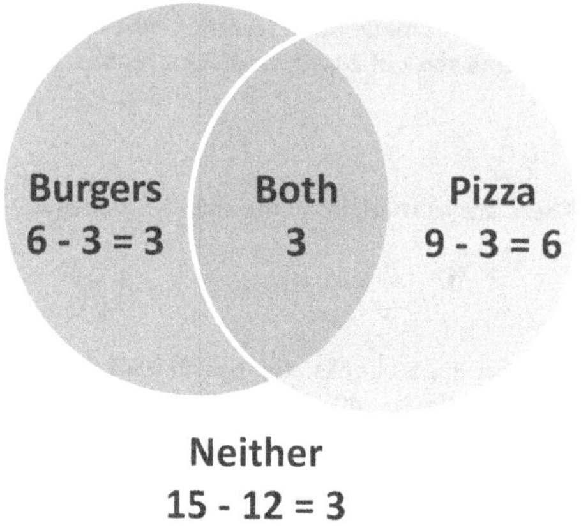

11) 9 - 2 x 3 + 10 =
Use the order of operations PEMDAS to solve.
9 - (2 x 3) + 10 =
9 - (6) + 10 = 13

12) 875 ÷ 25 = 35

13) $\frac{8}{12} \div \frac{2}{3} = \frac{8}{12} \times \frac{3}{2} = \frac{24}{24} = 1$

14) The average of these numbers: 15, 17, 19, 21, 23, 25, 27 is 21 because it is the middle number.

15) If Scott reads 2.1 pages per minute, how many pages does he read in 1.5 hours? 1.5 hours is 90 minutes. 90 x 2.1 = 189 pages.

PERENNIAL MATH

INDIVIDUAL ANSWER KEY
(Grades 3-4)

Instructions for Scoring:

- *Mark CORRECT answers by circling "O" the question number, and INCORRECT answers with an "X". Each correct answer is worth one point. Count the circles (correct answers) and write the number in the "SCORE" box.*

CORRECT → (1.) ________

INCORRECT → X. ________

1. 18
2. 13
3. 13
4. 50
5. 24
6. 19
7. (9, 17)
8. 4
9. 91
10. 3
11. 13
12. 35
13. 1
14. 21
15. 189

TEAM TESTS GRADES 3-4

20 minute practice tests for teams
Each member gets a copy of the test, but the team only receives one answer document

Grade 3-4 Team Event (20 Minutes)

1) What is the sum of all the odd numbers 1 through 35?

2) This tower is made of several cubes. Some cubes cannot be seen in the picture. How many cubes in all did it take to make this tower?

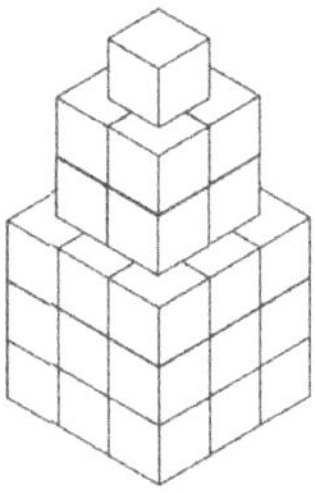

3) How many rectangles of any size are in this diagram?

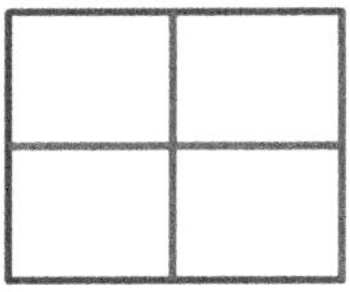

4) How many different coin combinations will give you 85 cents, if there are no pennies, and more dimes than nickels? Quarters and half dollars may also be used.

5) Meredith bought 4 pairs of pants, 3 shirts and 2 pairs of shoes. How many different outfits can she make?

6) The sum of 2 different numbers is 20. The difference of the two numbers is 6. What is the largest number of the two numbers?

7) When Micah counts her beads she notices that:
If she stacks them by twos she has one left over.
If she stacks them by threes she has one left over.
If she stacks them by fives she has none left over.
What is the least number of beads Micah could have?

8) Together Kennedy and Connie added their jellybeans and they had 60. Kennedy gave Connie 15 of her jellybeans and then the two girls had an equal amount. How many jellybeans did Connie have to start with?

9) Insects have 6 legs and spiders have 8. Bruno had spiders and insects in his box. If Bruno counted 72 legs altogether in the box, how many spiders were in his box?

10) Cameron averaged 11 points per game in the first 10 games, but he only scored 8 points in each of the next two games, how many total points did he score for all 12 games?

PERENNIAL MATH

TEAM ANSWER DOCUMENT

SCORE

(Grades 3-4)

Team # __________ School ______________________________

Student Names____

____

Write answers clearly. Each correct answer is worth one point.

1. __________
2. __________
3. __________
4. __________
5. __________
6. __________
7. __________
8. __________
9. __________
10. __________

Grade 3-4 Team Event Solutions

6) Make a list of numbers whose sum is 20 and look for 2 numbers whose difference is 6. The largest of the 2 numbers is 13.

		Sum	Difference
1	19	20	18
2	18	20	16
3	17	20	14
4	16	20	12
5	15	20	10
6	14	20	8
7	13	20	6
8	12	20	4
9	11	20	2

7) Make a chart to see what the lowest multiple of 5 is that is one more than a multiple of 2 and 3. The least amount of beads she can have is 25.

Multiples of 2 and 3		6	12	18	24
Multiples of 5	5	10	15	20	25

8) Kennedy + Connie = 60 jellybeans. When Kennedy gives Connie 15 jellybeans, they have the same amount. 60 jellybeans split in half would be 30 each to make the girls have an equal amount. So Connie received 15 jellybeans and she must have had 15 to start with because 15 + 15 = 30. If Kennedy gave away 15 jellybeans and has 30 now she had to have started with 45 because 45-15 =30 jelly beans now. So check your work :

Connie has 15 and Kennedy has 45. 15 + 45 = 60
Connie receives 15 so she has 15 + 15 = 30
Kennedy gives 15 away so she had 45 – 15 = 30

Answer: Connie had 15 to start with.

9) Bruno had spiders and insects in his box. If Bruno counted 72 legs altogether in the box, then use a table to see what combinations of legs will give you 72 total legs since there are 6 legs on insects and 8 legs on spiders. There were 6 spiders in the box.

	Number of legs per Critter	Number in the box	Number of legs
Insects	6	4	24
Spiders	8	6	48
Total			72

10) Cameron averaged 11 points per game in the first 10 games, so 11 x 10 = 110 points the first 10 games. He scored 8 points the next two games, which 8 x 2 = 16 additional points. Therefore, 110 + 16 = 126 points.

Grade 3-4 Team Event Solutions

1) The sum of all the odd numbers 1-35 is 1 + 3 + 5 + 7 + 9 + 11 + 13 +15 + 17 + 19 + 21 + 23 + 25 + 27 + 29 + 31 + 33 + 35 = 324

Note: the beginning and ending numbers add up to 36. For example 1 + 35, 3 + 33, 5 + 31, etc. So there are 9 pairs of 36 and 9 x 36= 324.

2) This tower is made up of several cubes.

- The bottom part is composed of a 3 x 3 x 3 or 27 cubes.
- The middle section is composed of a 2 x 2 x 2 or 8 cubes.
- The top part is a 1 x 1 or 1 cube.
- So there are 27 + 8 + 1 = 36 cubes in this tower.

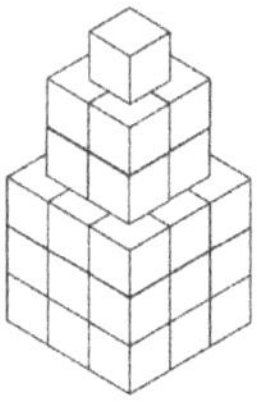

3) How many rectangles of any size are in this diagram?

Type	How Many
Made of 1 shape	4
Made of 2 shapes	4
Made of 3 shapes	0
Made of 4 shapes	1
Total rectangles	9

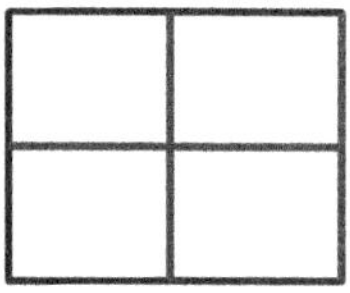

4) Make a chart to see the different coin combinations for 85 cents that uses half dollars, quarters, dimes and nickels and more dimes than nickels.

Half Dollar	Quarter	Dime	Nickel	Total
1	1	1	0	0.85
1	0	3	1	0.85
0	3	1	0	0.85
0	2	3	1	0.85
0	1	6	0	0.85
0	1	5	2	0.85
0	0	8	1	0.85
0	0	7	3	0.85
0	0	6	5	0.85

5) Meredith bought 4 pairs of pants, 3 shirts and 2 pairs of shoes. How many different outfits can she make?

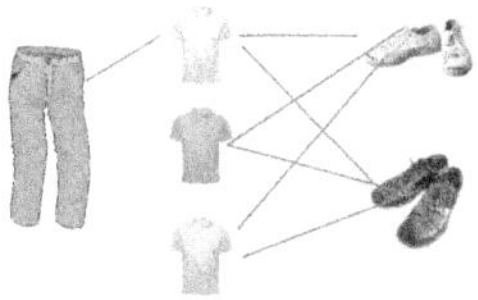

6 outfits for each pair of pants x 4 pairs of pants = 24 outfits.

Use multiplication principle:
4 x 3 x 2 = 24

TEAM ANSWER KEY

(Grades 3-4)

Mark answers incorrect. Each correct answer is worth one point. Write the number that the student got correct in the SCORE box on their document.

1. **324**
2. **36**
3. **9**
4. **9**
5. **24**
6. **13**
7. **25**
8. **15**
9. **6**
10. **126**

Grade 3-4 Team Event (20 Minutes)

1) The rain was falling 1 and ½ inches per hour. At this rate, how long would it take for 3 feet of rain to fall? *There are 12 inches in 1 foot.*

2) The perimeter of this large square is 40 cm. What is the perimeter of one of the smaller squares?

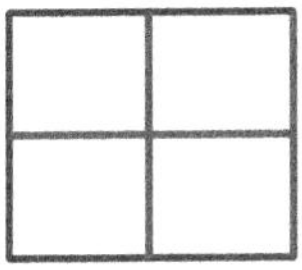

3) The sums of each row and column are given. How much is 1 triangle worth?

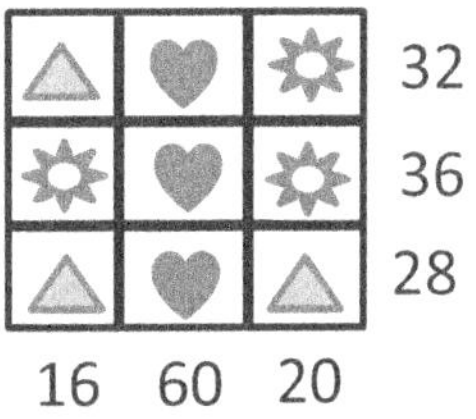

4) Carmen worked at the aquarium shop for 12 weeks this summer and got paid the same amount each week. She made a total of $1800. How much did she make each month?(*1 month = 4 weeks*)

5) A box of blocks can be divided equally among 3, 5, or 6 students. What is the least number of blocks in the box if there are more than 50?

6) Kyle has a black, blue, red, and white shirt. He has black, blue, and red pants. How many shirt and pants out fits can he make if he cannot wear the same color shirt and pants together?

7) Landon is digging worms for fishing. He finds 15 per minute, but 2 die every 15 seconds. How many minutes will it take him to have a total of 210 live worms ready to fish with?

8) A spinner has 15 sections. Six spaces are red, three are green, and six are yellow. What is the probability that you will land on a red space? Reduce to lowest terms.

9) Molly bought a pair of shoes. The original price was $40, but the shoes were on sale for 15% off. How much did she have to pay for the shoes?

10) Mike had to write his numbers down in order as fast as he could starting with 1 and going to 100. If he could only write 1 digit every second, what digit did he write last at the end of 1 minute?

 PERENNIAL MATH

TEAM ANSWER DOCUMENT

(Grades 3-4)

SCORE

Team # ___________ School ______________________________________

Student Names____

____

Write answers clearly. Each correct answer is worth one point.

1. ____________
2. ____________
3. ____________
4. ____________
5. ____________
6. ____________
7. ____________
8. ____________
9. ____________
10. ____________

Grade 3-4 Team Event Solutions

1) Divide 36 inches by 1.5 inches per hour and it takes 24 hours for 36 inches or 3 feet of rain to fall.

2) Since the perimeter of the large rectangle is 40cm, each side is 10 cm. The sides of the small squares are all equal to 5cm. The perimeter of one of the smaller squares is 4 sides x 5cm = 20 cm.

3) Since 3 hearts = 60, then 1 heart = 20. 2 triangles + 1 heart = 28. Subtract the heart(20) from the equation and the 2 triangles are 8. One triangle = 4.

4) Since we are figuring the pay for a month divide the total by 3. $1800 ÷ 3 = $600.

5) The least common multiple of 3, 5, and 6 is 30. Since there must be more than 50 blocks in the box, the lowest number that is divisible by 3, 5, and 6 is 60 blocks.

6) He could make 4 x 3 = 12 total outfits. However, he must subtract the three shirt/pant combinations of the same color black/black, blue/blue, and red/red. He can make 9 outfits.

7) He is digging 15 worms each minute, but 2 are being subtracted every 15 seconds. So every minute 8 are subtracted. That means he is only gaining 7 worms each minute. At that rate it will take him 210/7 = 30 minutes to dig the 210 worms.

8) There are 15 total sections. Since 6 are red, the probability of landing on red is 6/15 or 2/5.

9) To find out how much she saved multiply $40 x .15 = $6. Subtract to find out how much she paid for the shoes $40 - $6 = $34.

10) 1, 2, 3, 4, 5, 6, 7, 8, 9, 10, 11,12, 13, 14, 15, 16, 17, 18, 19, 20, 21, 22, 23, 24, 25, 26, 27, 28, 29, 30, 31, 32, 33, 34, 35 *This is the 60th digit*

TEAM ANSWER KEY

(Grades 3-4)

Instructions for Scoring:

- *Mark CORRECT answers by circling "O" the question number, and INCORRECT answers with an "X" . Each correct answer is worth one point. Count the circles (correct answers) and write that number in the "SCORE" box.*

1. **24 hours**
2. **20cm**
3. **4**
4. **$600**
5. **60**
6. **9**
7. **30 minutes**
8. **2/5 (6/15 is incorrect)**
9. **$34**
10. **3**

Grade 3-4 Team Event (20 Minutes)

1) Mrs. Williams turned 89 in 2016, what year was she born?

2) The numbers 5, 6, 7, and 8 can be placed in the empty squares so that each row, column, and diagonal contains each number only once. What will be the sum of the numbers placed in squares X and Y?

5		8	X
6			Y
7			
8		5	

3) Circles A, B, and C form four regions where they intersect. The numbers 1, 3, 4, and 6 can be placed in those regions so that each of the circles has the same sum. What is that sum?

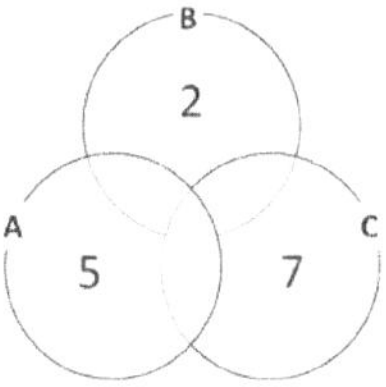

4) The four-digit number 7C3C is divisible by 9. What digit does the C represent?

5) A light flashes 4 times every 5 seconds. At this rate, how many times will the light flash in 5 minutes?

6) Kelsie has 10 bells. They have weights of 2g, 4g, or 5g. She has a different number of bells of each weight. What is the largest possible weight or total grams of Kelsie's bells?

7) Two years ago my age was a multiple of 7. Last year it was a multiple of 5. I am younger than 50. How old am I now?

8) Mary and Clair each have the same amount of tickets. Mary buys 1 candy bar with his tickets and still has 90 tickets left. Clair buys 4 candy bars and has 30 tickets left. How many tickets did Mary start with?

9) Mindy went shopping for a pillow, handed the clerk $30 and received $7 change. Next she bought a scarf, handed the clerk $15 and received $4 change. She had $22 in her wallet after shopping. How many dollars did she have in her wallet after she bought the pillow?

10) The farmer counted the cows and chickens in the pen and there were 25 heads and 64 feet. How many chickens were in the pen?

PERENNIAL MATH

TEAM ANSWER DOCUMENT

SCORE

(Grades 3-4)

Team # ___________ School ______________________________________

Student Names____

____

Write answers clearly. Each correct answer is worth one point.

1. ____________

2. ____________

3. ____________

4. ____________

5. ____________

6. ____________

7. ____________

8. ____________

9. ____________

10. ____________

Grade 3-4 Team Event Solutions

1) 2016-89 = 1927. Mrs. Williams was born in 1927.

2) Numbers 6 & 7 are missing in the 3rd column-The 7 must be in the 2nd row because there is already a 6, and also the 6 must be in the 3rd row because there is already a 7. Next the diagonal (top-left to bottom-right) can be filled in with the missing numbers 7 & 8 and the 8 must go in the 2nd row because there is already a 7. That will make Y have to be 5 and X would have to be 6 since the 1st row is missing 6 and 7 and there is already a 7 in that column.
The sum of X + Y or 5 + 6 = 11

5		8	X
6	8	7	Y
7		6	
8		5	7

5	7	8	6
6	8	7	5
7	5	6	8
8	6	5	7

3) The numbers 1, 3, 4, and 6 can be placed in the four regions where Circles A, B, and C intersect so that all circles have the same sum of 15.

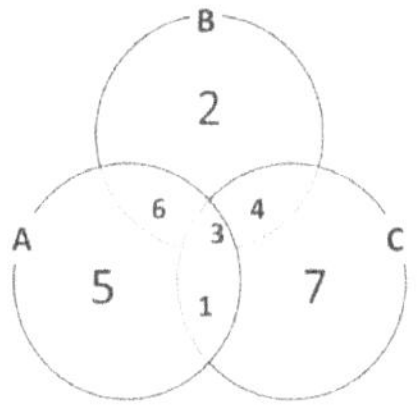

4) In order for a number to be divisible by 9, the sum of its digits must be divisible by 9. Since the sum of the digits of 7C3C is 10, then the next multiple of 9 is 18. Therefore, C + C must add up to 8 to make the full sum of all the digits be 18. Therefore, C must be 4.

5) Each minute has 60 seconds. So if the light flashes every 5 seconds, there are 12 flashing intervals each minute. If the light flashes 4 times each interval then there are 12 x 4, or 48 flashes per minute. The light flashes 48 x 5 or 240 flashes in 5 minutes.

6) Start with the most of amount of the heaviest bells and add the most amount of the next highest weight, but make sure that you have a different amount of balls. Kelsie has 7of the 5 gram bell, 2 of the 4 gram bells, and 1 of the 2 gram bells. 7(5) + 2(4) + 2 = 35 + 8 + 2 = 45 grams.

7) Create a table to get a perspective of numbers. Start by listing the multiples of 7's less than 50 across the table. Fill in the age for the next two rows and look for age that meets the requirements. Your age is 16.

Age 2 Years Ago (multiples of 7)	7	14	21	28	35	42
Age Last Year (look for multiples of 5)	8	15	22	29	36	43
Age Now	9	16	23	30	37	44

8) Clair bought 3 more candy bars than Mary and ended with 90 – 30 = 60 tickets less. So candy bars must cost 20 tickets each since 60 tickets/3 candy bars = 20. Since Mary still had 90 tickets and only bought one candy bar, then Mary started with 90 + 20 = 110 tickets .

9) Mindy spent (30-7) or $23 on the pillow. Then she also spent (15-4) or $11 on the scarf. She had $22 in her wallet when done shopping. $22 left + $34 spent = $56 when she started shopping. If she had $56 when she started shopping and spent $23 at the first store she would have had $33 in her wallet after she bought the pillow.

10) Make a table. Start with 12 chickens and 13 cows. Note that the total feet need to be reduced to get to 64, so start to increase number of chickens. There were 18 chickens in the pen.

Chickens	12	13	14	15	16	17	18
Cows	13	12	11	10	9	8	7
Chicken feet	24	26	28	30	32	34	36
Cow feet	52	48	44	40	36	32	28
Total feet	76	74	72	70	68	66	64

PERENNIAL MATH

TEAM ANSWER KEY

(Grades 3-4)

Instructions for Scoring:

- ***Mark CORRECT answers by circling "O" the question number, and INCORRECT answers with an "X". Each correct answer is worth one point. Count the circles (correct answers) and write that number in the "SCORE" box.***

CORRECT → (1.) ________

INCORRECT → X. ________

1. **1927**
2. **11**
3. **15**
4. **4**
5. **240**
6. **45**
7. **16**
8. **110**
9. **33**
10. **18**

Grade 3-4 Team Event (20 Minutes)

1) If I have 510 pennies, how many days will it take me to spend all of them if I spend 6 pennies a day?

2) How many dark squares will border a line of 13 white squares?

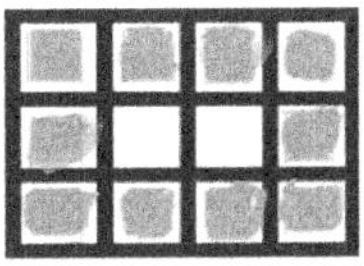

3) The perimeter of the rectangle is 48cm. The length is twice the size of the width. What is its area?

4) Mindy had 64 lollipops. She gave half of them to her teacher. She then gave an even number of them to her three best friends. She kept 5 of them for herself. How many did each of her best friends receive?

5) There are 25 students in the class. Fifteen students like chocolate, sixteen like vanilla, and 8 like both flavors. How many people like neither flavor?

6) My older sister is 6 years older than I am. If the sum of our ages is 32, how old is my sister?

7) There are 5 cars for every 3 trucks in the parking lot. If there are a total of 96 vehicles in the parking lot, how many trucks are there?

8) Amy has 3 different books arranged in line on her bookshelf. How many ways can she arrange those 3 books?

9) Mitchell had an equal number of quarters, dimes, nickels, and pennies in his piggy bank. He had more than $3.00. What is the least amount of money he could have had?

10) Old McDonald had chickens and cows in his barn. He counted 24 heads and 66 legs. How many cows did he have?

PERENNIAL MATH

TEAM ANSWER DOCUMENT

(Grades 3-4)

SCORE

Team # __________ School ______________________________________

Student Names____

____

Write answers clearly. Each correct answer is worth one point.

1. ____________

2. ____________

3. ____________

4. ____________

5. ____________

6. ____________

7. ____________

8. ____________

9. ____________

10. ____________

Grade 3-4 Team Event Solutions

1) If I have 510 pennies and I spend 6 pennies a day it will take me 510 ÷ 6 = 85 days to spend all the pennies.

2) How many squares will be in the dark border in a line of 13 white squares? The pattern for the border is 2 x the number of white squares plus the 3 squares on both ends (6 squares). So 13 white squares would have a border of (13 x 2) + 6 = 26 + 6 = 32 squares.

3) The perimeter of the rectangle is 48cm. The length is twice the size of the width. If the width of the rectangle is 8cm, then the length would be 16cm. So, the area of the rectangle would be 8 x 16 = 128cm^2.

4) Mindy had 64 lollipops. She gave half of them to her teacher. That would leave 32 lollipops. She then gave an even number of them to her three best friends. She kept 5 of them for herself. 32 – 5 = 27. So 27 ÷ 3 = 9 lollipops that she gave to her best friends.

5) Use a Venn Diagram to find how many students like neither. For chocolate 15 – 8 = 7 Chocolate only. For vanilla 16 – 8 = 8 Vanilla only. 7 + 8 + 8 = 23 student who like the flavors. 2 like Neither.

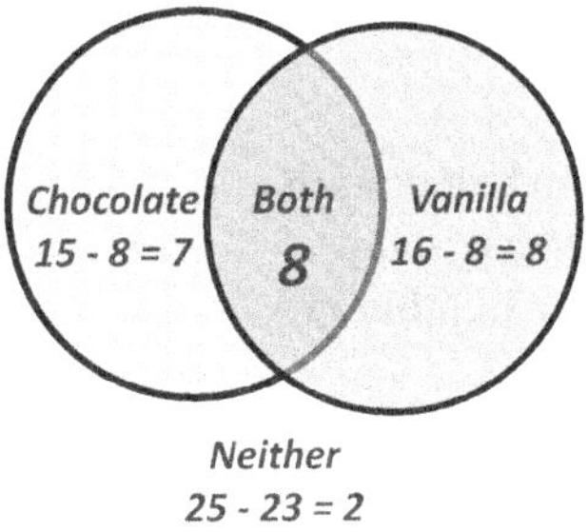

Grade 3-4 Team Event Solutions

6) Since my older sister is 6 years older than I am and the sum of our ages is 32, you can find how old my sister is with a table.

My Age	My Sister's Age	Sum
1	7	8
2	8	10
3	9	12
4	10	14
5	11	16
6	12	18
7	13	20
8	14	22
9	15	24
10	16	26
11	17	28
12	18	30
13	**19**	32

7) There are 5 cars for every 3 trucks in the parking lot. There are a total of 96 vehicles in the parking lot. To find out how many trucks are there divide 96 by 8 to get 12 groups of vehicles. So there are 12 x 3 = 36 trucks.

8) Amy has 3 different books arranged in line on her bookshelf. To find out how many ways can she arrange those 3 books make a chart.

Book 1 Book 2 Book 3
Book 1 Book 3 Book 2
Book 2 Book 1 Book 3
Book 2 Book 3 Book 1
Book 3 Book 1 Book 2
Book 3 Book 2 Book 1

There are 6 ways she can arrange the books.

9) Mitchell had an equal number of quarters, dimes, nickels, and pennies in his piggy bank. He had more than $3.00. Make a chart to find the least amount of money he could have had. He would have had $3.28

Quarters	Dimes	Nickels	Pennies	Sum
1	1	1	1	0.41
2	2	2	2	0.82
3	3	3	3	1.23
4	4	4	4	1.64
5	5	5	5	2.05
6	6	6	6	2.46
7	7	7	7	2.87
8	8	8	8	3.28

10) Use a chart and start with half of each animal. Notice that you have too many legs and need to decrease the number of cows. He had 9 cows.

Cows	12	11	10	9
Chickens	12	13	14	15
Total heads	24	24	24	24
Cow Feet	48	44	40	36
Chicken Feet	24	26	28	30
Total feet	72	70	68	66

PERENNIAL MATH

TEAM ANSWER KEY

(Grades 3-4)

Instructions for Scoring:

- ***Mark CORRECT answers by circling "O" the question number, and INCORRECT answers with an "X" . Each correct answer is worth one point. Count the circles (correct answers) and write that number in the "SCORE" box.***

CORRECT → (1.) ________

INCORRECT → ~~1.~~ ________

1. **85**
2. **32**
3. **128**
4. **9**
5. **2**
6. **19**
7. **36**
8. **6**
9. **3.28**
10. **9**